Cosmic ray and interplanetary magnetic field during the passage of Coronal mass ejections

Cosmic Rays and CMEs

Rajesh Kumar Mishra
Rekha Agarwal

Bibliographic information published by the German National Library:

The German National Library lists this publication in the National Bibliography; detailed bibliographic data are available on the Internet at http://dnb.dnb.de.

ISBN: 9783346979384
This book is also available as an ebook.

© GRIN Publishing GmbH
Trappentreustraße 1
80339 München

Print and binding: Books on Demand GmbH, Norderstedt, Germany
Printed on acid-free paper from responsible sources.

GRIN web shop: https://www.grin.com/document/1425394

Cosmic ray and interplanetary magnetic field during the passage of Coronal mass ejections

Rajesh K. Mishra [1] and Rekha Agarwal [2]

[1] Information Technology Cell, Tropical Forest Research Institute,

P.O.: RFRC, Mandla Road, Jabalpur (M.P.) India 482 021

[2] Department of Physics, Govt. Model Science College (Autonomous),

Jabalpur (M.P.) 482 001, India

Abstract

Coronal Mass Ejections are vast structures of plasma and magnetic fields that are expelled from the sun into the heliosphere. This material is detected by remote sensing and in-situ spacecraft observations. The present study deals with the influence of four types of CMEs namely Asymmetric 'Full' Halo CMEs, Partial Halo CMEs, Asymmetric and Complex 'Full' Halo CMEs and 'Full' Halo CMEs on cosmic ray neutron monitor intensity. The data of ground based neutron monitor of Moscow and CME events observed with instruments onboard and Wind spacecraft have been used in the present analysis. The method of superposed epoch (Chree) analysis has been used to the arrival times of these CMEs. It is noteworthy that the frequency of occurrence of Asymmetric 'Full' Halo CMEs is significantly high, whereas frequency of occurrence of Asymmetric and Complex 'Full' Halo CMEs is low compared to other CMEs. Significant enhancement in cosmic ray intensity is observed after 4 days of the onset of asymmetric full halo and 6 days after the onset of full halo CMEs. The fluctuations in cosmic ray intensity are more prior to the onset of both types of the CMEs. However, during Partial Halo CMEs the cosmic ray intensity peaks, 8- 9 days prior to the onset of CMEs and depressed 3 days prior to the onset of CMEs, whereas in case of asymmetric and complex full CMEs, the intensity depressed 2 days prior to the onset of CMEs and enhanced 2 days after the onset of CMEs. The deviations in cosmic ray intensity are more pronounced in case for asymmetric and complex full halo CMEs compared to other CMEs. The cosmic ray intensity shows nearly good anti-correlation with IMF strength (B) during asymmetric full halo CMEs and partial halo CMEs, whereas it shows poor correlation with B during other CMEs. The IMF, B significantly decreased five days prior to the onset of asymmetric and full halo CMEs and four days after the onset of partial halo CMEs, whereas IMF strength (B) significantly enhanced 5-6 days prior and after the onset of asymmetric and complex full halo CMEs. IMF strength (B) significantly depressed 2 days prior and 4 days after the onset of full halo CMEs. However, IMF, B significantly enhanced from its minimum to maximum values in 2 days interval prior to the onset of CMEs and in 3 days interval after the onset of CMEs

PACS Nos.: 96.40.Kk, 96.40. -z, 96.40.cd

Keywords: cosmic ray, interplanetary magnetic field, solar activity and coronal mass ejections

Introduction

Earlier, it was thought that solar flares were responsible for major interplanetary particle events and geomagnetic storms. However, recently we have seen an important paradigm shift such that now coronal mass ejections (CMEs), not flares, are considered the key causal link with solar activity. CMEs are plasma eruptions from the solar atmosphere involving previously closed field regions, which are expelled into the interplanetary medium. Such regions, and the shocks which they may generate, have pronounced effects on cosmic ray densities both locally and at some distance away. These energetic particle effects can often be used to identify CMEs in the interplanetary medium, where they are usually called `ejecta'. When both the ejecta and shock effects are present the resulting cosmic ray event is called a `classical, two-step' Forbush decrease.

Bieber and Evenson [1] noticed strong enhancements of the cosmic ray anisotropy before and during the January 1997 CME/magnetic cloud. From a multi-station analysis of neutron monitor data, they conclude that $\mathbf{B}\times\nabla n$ drift is a primary source of CME-related anisotropies for 5 GeV cosmic rays. Evolution of the cosmic ray density and density gradients is closely linked to magnetic properties of the ejecta, and provides information on the magnetic cloud and related features as they approach and pass Earth. Strong enhancement of the field-aligned anisotropy was observed primarily during the 9 hours prior to shock arrival condition of Earth. Cane et al. [2] reported a significant relationship between CMEs and cosmic ray variations.

Shrivastava [3] argued that the coronal mass ejections in association with B-type solar flare might be the reason for the enhancement of geomagnetic field variation and CMEs indicate its better role in cosmic ray modulation.

The intensity of galactic cosmic rays measured on Earth is related to the Sun's cycle of activity, which is well known by astronomers. The solar magnetic field flips every 11 years and the number of sunspots and 'coronal mass ejections' rises and falls twice in each complete 22-year cycle. The cosmic ray intensity on Earth also peaks twice every 22 years in time with the solar cycle. Cliver and Ling [4] have discovered a quirk in this pattern - and they believe that coronal mass ejections could be responsible for it.

Edward Cliver, of the Air Force Research Laboratory in Massachusetts, and Alan Ling, of Redex Inc in Massachusetts, compared numbers of sunspots - dark patches on the disk of the sun caused by local magnetic fields - and measurements of galactic cosmic rays dating back to 1951. The sharp fall in cosmic ray intensity that occurs every 11- year is closely related to the rise in the number of sunspots. They studied this relationship and noticed that the cosmic ray curve lagged behind the rise in the number of sunspots by about a year - but only during alternate solar cycles. In the intervening cycles, the two trends occur almost simultaneously.

The researchers suspect that the alternating pattern is rooted in the reversal of the Sun's magnetic field every 11 years. Cosmic rays preferentially approach the Sun from the direction of its poles when the magnetic field lines are pointing out of the Northern hemisphere. When the magnetic field flips, cosmic rays tend to approach equatorial regions of the Sun. But astronomers also know that coronal mass ejections (CMEs) - colossal streams of gas that erupt from the Sun's surface - tend to occur close to the Sun's equator early in the solar cycle, and later migrate towards the poles.

Cliver and Ling [4] propose that when cosmic rays impinge on the solar poles early in an 11-year cycle, they do not encounter CMEs. But cosmic rays do meet CMEs when they approach the equator at this time in the solar cycle. This means that the interaction of cosmic rays with the strong magnetic fields of CMEs affects the intensity of cosmic rays on Earth. There are many uncertainties inherent in predicting long-term trends from relatively short-term measurements, as Cliver and Ling point out. But the pattern is clearly evident from the data so far.

We present a study of the short-term evolution of coronal mass ejections observed by the Large Angle and Spectrometric Coronograph (LASCO) on board SOHO during 2005 and their association with the modulation of galactic cosmic ray (GCR) intensity observed at 1 AU by the Moscow neutron monitor and IMP-8 spacecraft. We compare the short-term GCR modulation with the CME occurrence rate at all, low, and high latitudes, as well as the observed CME parameters.

Data and analysis

CME events observed by instruments onboard SOHO and Wind space craft for the period 2005-06 have been considered for the present work. We have analyzed sixty-seven CMEs during 2005-06. The temperature and pressure corrected hourly data (counts of neutrons) of cosmic ray intensity from Moscow neutron super monitor 24NM-64 (Latitude 55.47 N, Longitude 37.32 E, Altitude 200 m, Standard pressure 1000 mb, Geomagnetic cut-off rigidity 2.43 GV) have been used, where the long-term change from the data has been removed by the method of trend correction. Chree analysis of superposed epoch has been applied on the presure corrected daily average cosmic ray intensity data with respect to full hallo CMEs, partial hallo CMEs, Asymmetric and Complex 'Full' Halo CMEs and asymmetric hallo CMEs. Statistical significance of the results so obtained is evaluated by using a method suitable for Chree analysis.

Results and Discussion

We have selected CMEs and divided in to four groups (1) Asymmetric 'Full' Halo CMEs, (2) Partial Halo CMEs (3) Asymmetric and Complex 'Full' Halo CMEs and (4) 'Full' Halo CMEs during 2005-06. We have adopted the Chree analysis of superposed epoch to study the effect of these CMEs on cosmic ray intensity using the daily average cosmic ray intensity of Moscow neutron monitor during 2005-06.

Fig 1 shows the frequency of occurrence of four different types of CMEs identified during the period 2005-06. It is clearly seen from the figure that frequency of occurrence of Asymmetric 'Full' Halo CMEs is significantly high, whereas frequency of occurrence of Asymmetric and Complex 'Full' Halo CMEs is low compared to other CMEs identified during the period of investigation. It is also noticed that frequency of occurrence of full halo and partial halo CMEs is almost equal.

To study the effect of these CMEs on cosmic ray intensity, we have adopted the Chree analysis of superposed epoch for days – 10 to + 10 and plotted in Fig 2 (a, b, c, d) as a percent deviation of cosmic ray intensity data of Moscow neutron super monitor for 2005-06. Deviation for each event is obtained from the overall average of 21 days. Epoch day (zero day) correspond to the starting days of CMEs. As depicted in Fig 2a the decrease in cosmic ray intensity on the onset of asymmetric full halo CMEs starts from – 10 day and reaches its first minimum on – 8 day, recovered on – 5 day and then decreases and reaches its second minimum on -3 day. Significant increase in cosmic ray intensity starts 3 days prior to the onset of CMEs and reaches its maximum 4 days after the onset of CMEs. The intensity then gradually decreases with some deviations up to + 9 day. It is also noticed from the plot that cosmic ray intensity significantly enhanced 4 days after the onset of the CMEs, whereas it fluctuates in 2-3 days interval prior to the onset of CMEs.

As depicted in Fig 2b during partial halo CMEs the increase in cosmic ray intensity starts from – 10 day and continues up to –8 day (first maximum). Significant decrease in cosmic ray intensity starts 8 days prior to the onset of CMEs and reaches its minimum 3 days prior to the onset of CMEs. The cosmic ray intensity then increases sharply up to zero epoch days. As seen in the plot the intensity remains statistically constant for few days (3 days) after the onset of CMEs. The cosmic ray intensity increases 3 days after the onset of CMEs and reaches its second maximum on + 5 day the start decreasing and reaches its previous value on +7 day. The intensity then increases and reaches its third maximum in one-day interval on +8 day and then decreases and reaches its usual value on +10 day. It is also observed from the plot that cosmic ray intensity significantly enhanced 8 days prior and decreased 3 days prior to the onset of the CMEs, whereas it fluctuates in 2-3 days interval 3 days after the onset of CMEs

Fig 2c shows that during asymmetric and complex 'full' halo CMEs, significant decrease in cosmic ray intensity starts from – 10 day and reaches its first minimum 2 days prior to the onset of CMEs. The intensity then significantly increases and reaches its first maximum 2 days after the onset of CMEs. Gradual decrease in cosmic ray intensity is seen, which starts 2 days after the onset of CMEs and continues up to 7 days after the onset of CMEs and reaches its second minimum. The intensity then increases up to + 9 day and reaches its second maximum. One of the significant observations during the onset of asymmetric and complex 'full' halo CMEs is that cosmic ray intensity significantly decreased 2 days prior and enhanced 2 days after the onset of CMEs. However, the deviation in the cosmic ray intensity is more after the onset of CMEs.

As shown in Fig 2d during full halo CMEs the deviations in cosmic ray intensity is very slow from -10 day to –5 day. The intensity is seems to be statistically constant for this period. The cosmic ray intensity then increases sharply with some fluctuations and reaches its first maximum, two days prior to the onset of CMEs. The intensity then decreases up to +1 day. As shown in the plot, gradual increases in cosmic ray intensity start, one day after the onset of CMEs and reaches its second maximum with some deviations 6 days after the onset of CMEs. The intensity then decreases significantly up to +10 day and reaches its usual value (i.e. as on epoch day). It is clearly seen from the plot that the cosmic ray intensity significant enhanced 6 days after the onset of full halo CMEs and fluctuations in cosmic ray intensity is more prior to the onset of CMEs.

Badruddin and Singh [6] studied the influence of CMEs; halo CMEs and partial halo CMEs on cosmic ray intensity and modulators as compared to the other CMEs. Lara et al. (2005) [7] studied the long-term evolution of CMEs observed by LASCO on board *SOHO* during the ascending, maximum, and part of the descending phases of solar cycle 23 and their relation with the modulation of galactic cosmic-ray (GCR) intensity observed at 1 AU by the Climax neutron monitor and *IMP-8* spacecraft. They observed a general anti-correlation between GCR intensity and the CME rate, which is relatively high (~-0.88), a lower anti-correlation between the low-latitude the CME rate and GCR intensity (~-0.71) and a very high anti-correlation between the high-latitude CME rate and GCR intensity (~-0.94). Their results suggest that all CME properties show some correlation with the GCR intensity, although there is no specific property (width, speed, or a proxy of energy) that definitely has a higher correlation with GCR intensity.

Significant enhancement in cosmic ray intensity is evident after 4 days of the onset of asymmetric full halo and 6 days after the onset of full halo CMEs. The fluctuations in cosmic ray intensity are more prior to the onset of both types of the CMEs. However, in case of Partial Halo CMEs the cosmic ray intensity peaks, 8- 9 days prior to the onset of CMEs and depressed 3 days prior to the onset of CMEs. In case of asymmetric and complex full CMEs, the intensity of cosmic rays depressed 2 days prior to the onset of CMEs and enhanced 2 days after the onset of CMEs. Thus, we may conclude that all the four types of CMEs studied here produced significant disturbances in cosmic ray intensity. However, the deviations in cosmic ray intensity are more pronounced in case for asymmetric and complex full halo CMEs. Short term modulation in cosmic ray intensity are caused by interplanetary shocks, which are driven by matter that is expelled from the Sun during a reorganization of the solar magnetic field i.e. CMEs. Most of CMEs are related with a specific solar flare and generate an interplanetary shock. The ejecta known to be the driver of interplanetary shocks. Magnetic cloud is also investigated as ejecta. These ejecta have a magnetic enhancement, which shows a clear rotation in the field direction. The CMEs have considerable influence on particle propagation and the interaction of these flows with quite solar wind create regions of compressed, heated solar wind and shocks, which are responsible for the modulation of cosmic rays.

Bothmer et. al. [8] studied the effect of high latitude CMEs on GCRs using measurement from the Kiel electron telescope. They find the passage of these CMEs over the spacecraft to be associated with the short-term decreases of GCR intensities. The relatively weak shocks in these events, driven by the CME's over-expansion, had no strong influence on the GCRs. Variation in the cosmic ray density has been investigated during an exceptional full halo CME on June 6, 2000 by Kandemir et al. [9]. They observed that the cosmic ray density has dropped drastically starting on June 8, 2000. During strong solar modulation, the local counts of cosmic ray intensity values dropped down as much as 24%.

Further to investigate the effect of these CMEs on interplanetary magnetic field (IMF) strength B, we have applied the Chree analysis of superposed epoch for days – 10 to + 10 and plotted in Fig 3 (a, b, c, d) as a percent deviation of IMF, B data for 2005-06.

One can clearly see for the plot 3a that IMF, B significantly increased 10 days prior to the onset of asymmetric full halo CMEs and reaches its highest on day –9. The IMF strength, B then decreased with deviations and reaches its minimum 5 days prior to the onset of CME. The IMF, B then increases gradually with slight deviations up to the epoch day i.e. on the arrival of CMEs. It is also noticed that IMF, B remains statistically constant one day prior and after the onset of these CMEs. The IMF, B fluctuates frequently in 1-2 day interval after the onset of asymmetric full halo CMEs and reaches its second maximum on day –10.

As depicted in Fig 3b, fluctuations in IMF strength (B) are quite frequent in 2-3 day interval prior to the onset of partial halo CMEs, which continues up to +1 day. A significant decrease in IMF, B seen starting one day after the onset of CMEs and it reaches its lowest on +4 day. The IMF, B then increases and recovered to its previous value on day +6 and then continues to fluctuate up to day +10.

The IMF strength (B) as depicted in Fig 3c fluctuates in 2 days interval from day –10 to day – 6 and then remains statistically constant for few days (i. e. from day –6 to day –4). The IMF B then start decreasing 4 days prior to the onset of asymmetric and complex full halo CMEs and reaches its minimum on zero epoch day with slight deviation on day –2. The increase in IMF, B is seen starting from zero epoch days and gradually it reaches its maximum 7 days after the onset of these CMEs with some deviations. The IMF strength then fluctuates in 2 days interval up to day –10. It is quite observed from this plot that

IMF strength (B) significantly enhanced 5-6 days prior and after the onset of asymmetric and complex full halo CMEs.

During full halo CMEs as depicted in Fig 3 d, the IMF strength (B) start decreasing 6 days prior to the onset of these CMEs and reaches its minimum on day –2. The IMF, B significantly enhanced in 2 days interval from day –2 to day –1. One day after the onset of full halo CMEs, the IMF, B decreases sharply with slight deviations and reaches its second minimum 4 days after the onset of CMEs. The IMF, B then again significantly enhanced in 3 days interval from day + 4 to day +6 and then fluctuates up to day +10. Here it is noteworthy that IMF strength (B) significantly depressed 2 days prior and 4 days after the onset of full halo CMEs. However, IMF, B significantly enhanced from its minimum to maximum values in 2 days interval prior to the onset of CMEs and in 3 days interval after the onset of CMEs.

We have also plotted the scattered diagram between cosmic ray intensity and IMF strength (B) along with regression line, correlation coefficient (r) to find out the possible correlation between these parameters during the onset of four different types of CMEs in Fig 4 (a, b, c, d). The correlation coefficient between these two parameters is also calculated. As depicted in the plots and observed correlation coefficient, the cosmic ray intensity shows nearly good anti-correlation with IMF strength (B) during asymmetric full halo CMEs (r = - 0.44) and partial halo CMEs (r = - 0.46), whereas a poor correlation is seen between these two during asymmetric and complex full halo CMEs (r = - 0.14) and full halo CMEs (r = 0.24).

Conclusions

From the present investigations following conclusions may be drawn:

The frequency of occurrence of Asymmetric 'Full' Halo CMEs is significantly high, whereas frequency of occurrence of Asymmetric and Complex 'Full' Halo CMEs is low compared to other CMEs.

Significant enhancement in cosmic ray intensity is observed after 4 days of the onset of asymmetric full halo and 6 days after the onset of full halo CMEs. The fluctuations in cosmic ray intensity are more prior to the onset of both types of the CMEs.

During Partial Halo CMEs the cosmic ray intensity peaks, 8- 9 days prior to the onset of CMEs and depressed 3 days prior to the onset of CMEs, whereas in case of asymmetric and complex full CMEs, the intensity depressed 2 days prior to the onset of CMEs and enhanced 2 days after the onset of CMEs.

The deviations in cosmic ray intensity are more pronounced in case for asymmetric and complex full halo CMEs compared to other CMEs.

The IMF, B significantly decreased five days prior to the onset of asymmetric and full halo CMEs and four days after the onset of partial halo CMEs, whereas IMF strength (B) significantly enhanced 5-6 days prior and after the onset of asymmetric and complex full halo CMEs

IMF strength (B) significantly depressed 2 days prior and 4 days after the onset of full halo CMEs. However, IMF, B significantly enhanced from its minimum to maximum values in 2 days interval prior to the onset of CMEs and in 3 days interval after the onset of CMEs

The cosmic ray intensity shows nearly good anti-correlation with IMF strength (B) during asymmetric full halo CMEs and partial halo CMEs, whereas it shows poor correlation with B during other CMEs.

Thus, we can say that CMEs are more effective transient modulators of cosmic ray intensity. However, study of the simultaneous deviations in solar wind plasma field parameters during the passage of these CMEs, their transit speed, magnetic field enhancements etc. are needs to be studied in more detail for establishment a better model.

Acknowledgements

The authors are indebted to various experimental groups, in particular, Prof. Margret D. Wilson, Prof. K. Nagashima, Miss. Aoi Inoue and Prof. J. H. King for providing the data. The SOHO/LASCO data used here are produced by a consortium of the Naval Research Laboratory (USA), Max-Planck-Institute fuer Aeronomie (Germany)), Laboratoire d'Astronomie (France), and the University of Birmingham (UK). SOHO is a project of international cooperation between ESA and NASA.

References

[1] Bieber, J. W., Evenson, Paul, 1998. CME geometry in relation to cosmic ray anisotropy, Geophys. Res. Lett., 25 (15), 2955-2958.

[2] Cane, H. V., Richardson, I.G., Von Rosenvinge, T.T., 1996. Cosmic ray decreases 1964-1994, J. Geophys. Res., 101, 21561.

[3] Shrivastava, P. K., 2001. Study of coronal mass ejections in relation with geomagnetic activity and cosmic rays, Proc. 27th Int. Cosmic Ray Conf., 8, 3425 - 3428.

[4] Cliver, E. W., Ling, A. G., 2001. Coronal mass ejections, open magnetic flux and cosmic ray modulation, The Astrophysical Journal, 556 (1), 432-437.

[5] Bieber, J. W., Evenson, P., 1997. Proc. 24th Int. Cos. ray Conf., 4, 341-344.

[6] Badruddin, Singh, Y. P., 2005. CME types, their interplanetary manifestations and effects on cosmic ray intensity, Proc. 29th Int. Cos. ray Conf., 3631-3634.

[7] Lara, A., Gopalswamy, N., Caballero-López, R. A., Yashiro, S., Xie, H., Valdés-Galicia, J. F., 2005. Coronal mass ejections and galactic cosmic ray modulation, The Astrophysical Journal, 625 (1), 441- 450.

[8] Bothmer, V., Heber, B., Kunow, H., Mueller-Mellin, R., Wibberenz, G., 1997. Effects of coronal mass ejection on galactic cosmic rays in the high latitude heliosphere: Observations from Ulysse's first orbit, Technical Report, Cosmic ray Assistant Secretary for Management and Administration.

[9] Kandemir, G., Geckinli, M., Firat, C., Yilmaz, M., Ozugur, B., 2002. Cosmic ray intensity variation during a CME, Planetary and Space Science, 50 (5-6), 633-636.

Caption to Figures

Fig 1: Frequency of occurrence of (1) Asymmetric 'Full' Halo CMEs, (2) Partial Halo CMEs (3) Asymmetric and Complex 'Full' Halo CMEs and (4) 'Full' Halo CMEs during 2005.

Fig 2: The results of Chree analysis of superposed epoch from –10 to +10 days with respect to zero epoch days as a percent deviation of cosmic ray intensity for (a) Asymmetric 'Full' Halo CMEs, (b) Partial Halo CMEs, (c) Asymmetric and Complex 'Full' Halo CMEs and (d) 'Full' Halo CMEs.

Fig 3: The results of Chree analysis of superposed epoch from –10 to +10 days with respect to zero epoch days as a percent deviation of interplanetary magnetic field strength for (a) Asymmetric 'Full' Halo CMEs, (b) Partial Halo CMEs, (c) Asymmetric and Complex 'Full' Halo CMEs and (d) 'Full' Halo CMEs.

Fig 4: Cosmic ray intensity along with the variation in associated value interplanetary magnetic field strength (B), regression line and correlation coefficient (r) for (a) Asymmetric 'Full' Halo CMEs, (b) Partial Halo CMEs, (c) Asymmetric and Complex 'Full' Halo CMEs and (d) 'Full' Halo CMEs.

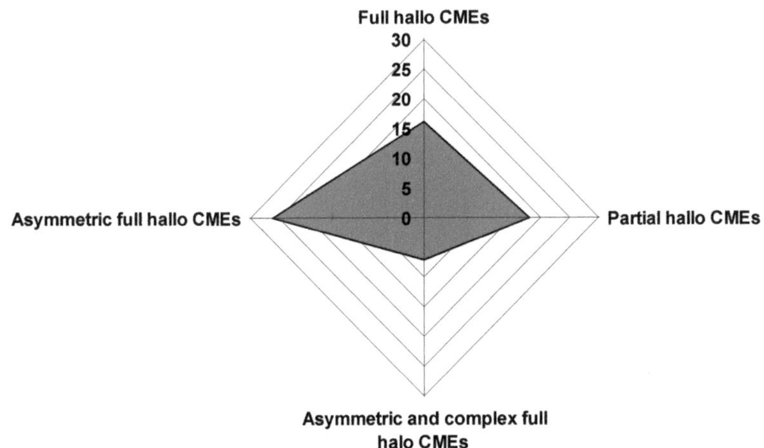
Full hallo CMEs
30
25
20
15
10
5
0
Partial hallo CMEs
Asymmetric and complex full halo CMEs
Asymmetric full hallo CMEs

FIG 1

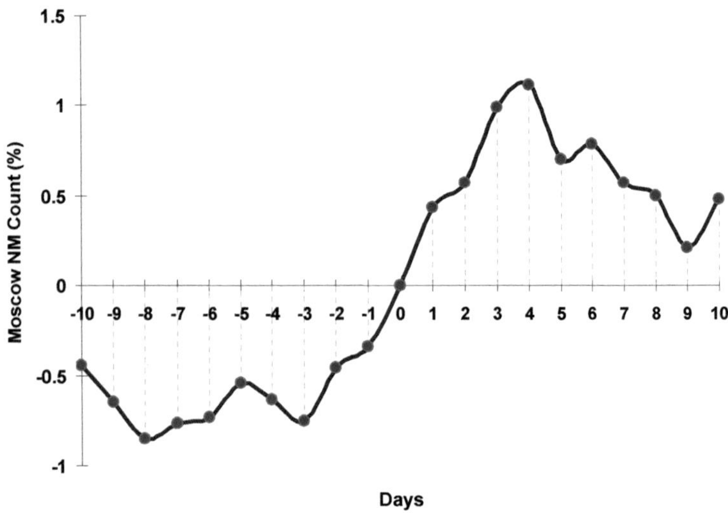
Asymmetric 'Full' Halo CMEs
1.5
1
0.5
0
-0.5
-1
Moscow NM Count (%)
-10 -9 -8 -7 -6 -5 -4 -3 -2 -1 0 1 2 3 4 5 6 7 8 9 10
Days

FIG 2a

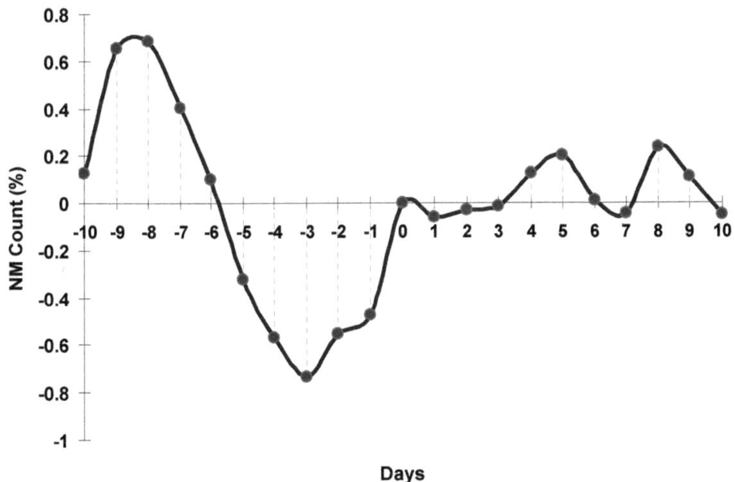
Partial Halo CMEs
0.8
0.6
0.4
0.2
0
-0.2
-0.4
-0.6
-0.8
-1
NM Count (%)
-10 -9 -8 -7 -6 -5 -4 -3 -2 -1 0 1 2 3 4 5 6 7 8 9 10
Days

Fig 2b

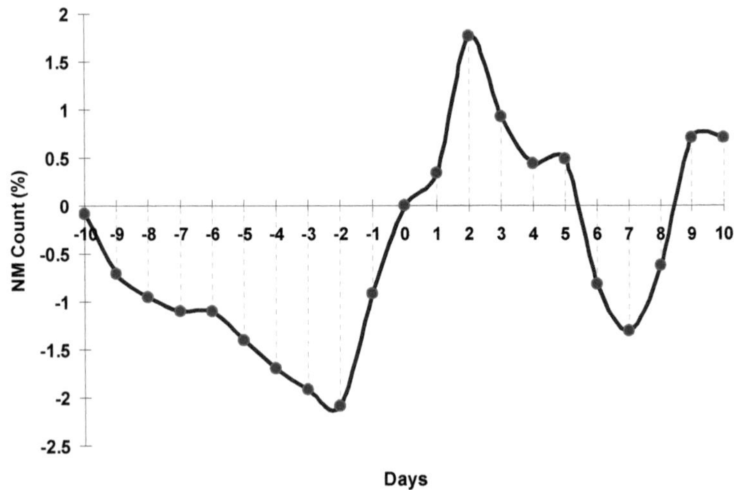
Asymmetric and Complex 'Full' Halo CMEs
2
1.5
1
0.5
0
-0.5
-1
-1.5
-2
-2.5
NM Count (%)
-10 -9 -8 -7 -6 -5 -4 -3 -2 -1 0 1 2 3 4 5 6 7 8 9 10
Days

Fig 2c

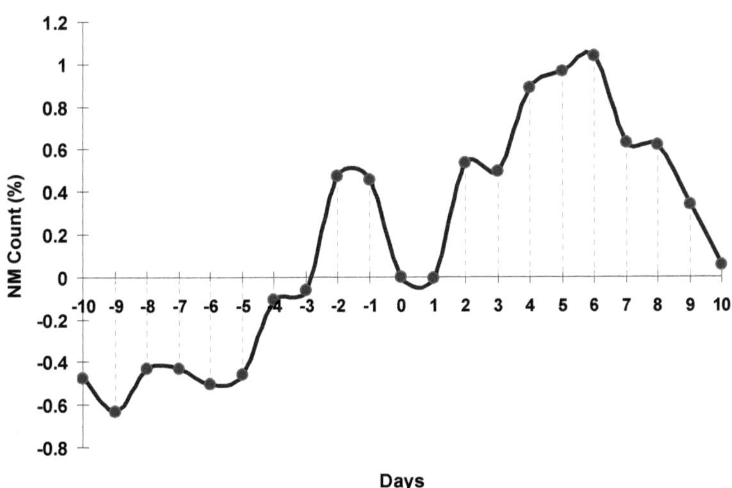
'Full' Halo CMEs
NM Count (%)
1.2
1
0.8
0.6
0.4
0.2
0
-0.2
-0.4
-0.6
-0.8
-10 -9 -8 -7 -6 -5 -4 -3 -2 -1 0 1 2 3 4 5 6 7 8 9 10
Days

Fig 2d

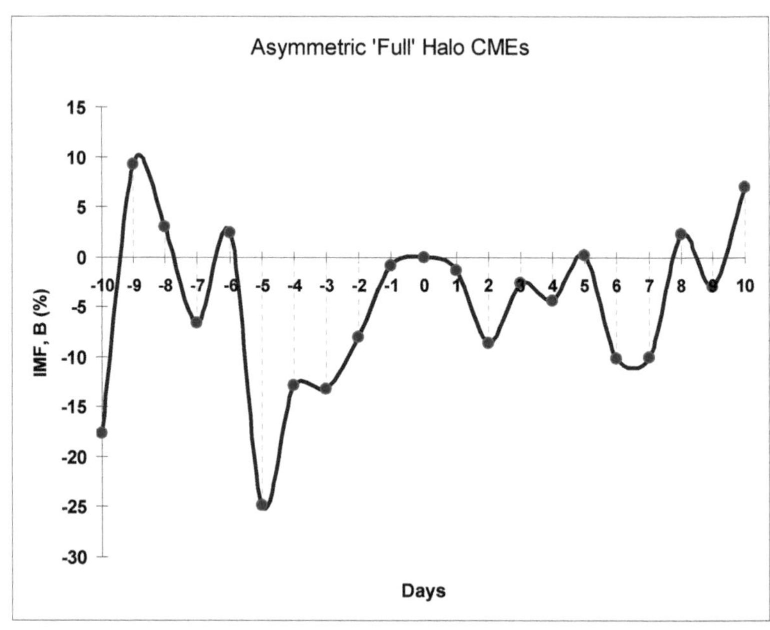

Asymmetric 'Full' Halo CMEs
15
10
5
0
-5
-10
-15
-20
-25
-30
IMF, B (%)
-10 -9 -8 -7 -6 -5 -4 -3 -2 -1 0 1 2 3 4 5 6 7 8 9 10
Days

Fig 3a

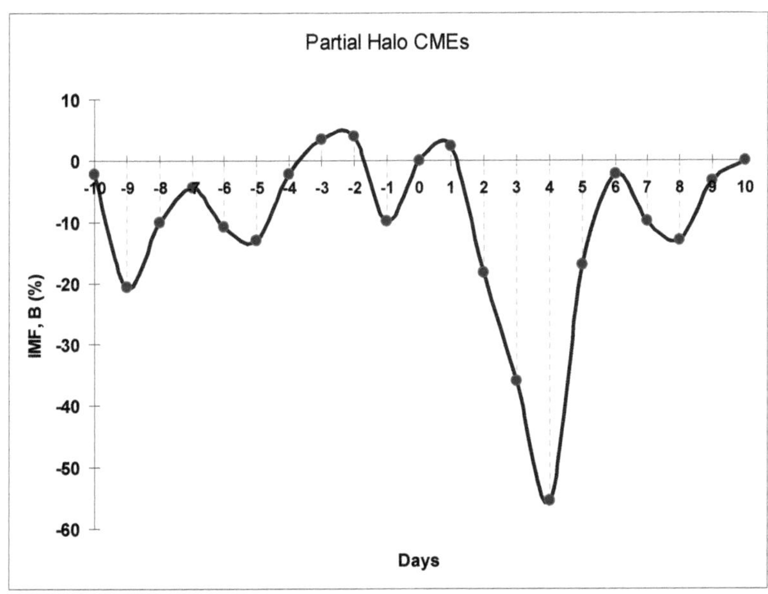
Partial Halo CMEs
10
0
-10
-20
-30
-40
-50
-60
IMF, B (%)
-10 -9 -8 -7 -6 -5 -4 -3 -2 -1 0 1 2 3 4 5 6 7 8 9 10
Days

Fig 3b

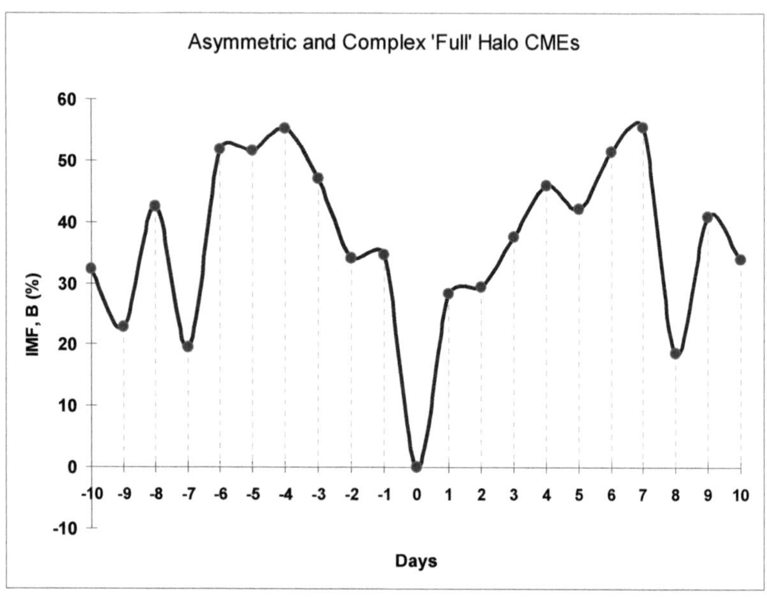
Asymmetric and Complex 'Full' Halo CMEs
IMF, B (%)
60
50
40
30
20
10
0
-10
-10 -9 -8 -7 -6 -5 -4 -3 -2 -1 0 1 2 3 4 5 6 7 8 9 10
Days

Fig 3 c

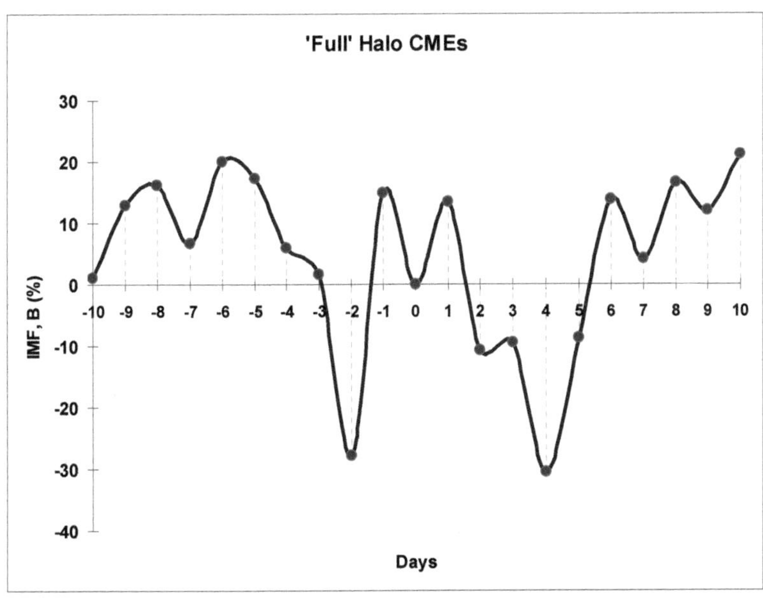
'Full' Halo CMEs
30
20
10
0
-10
-20
-30
-40
IMF, B (%)
-10 -9 -8 -7 -6 -5 -4 -3 -2 -1 0 1 2 3 4 5 6 7 8 9 10
Days

Fig 3d

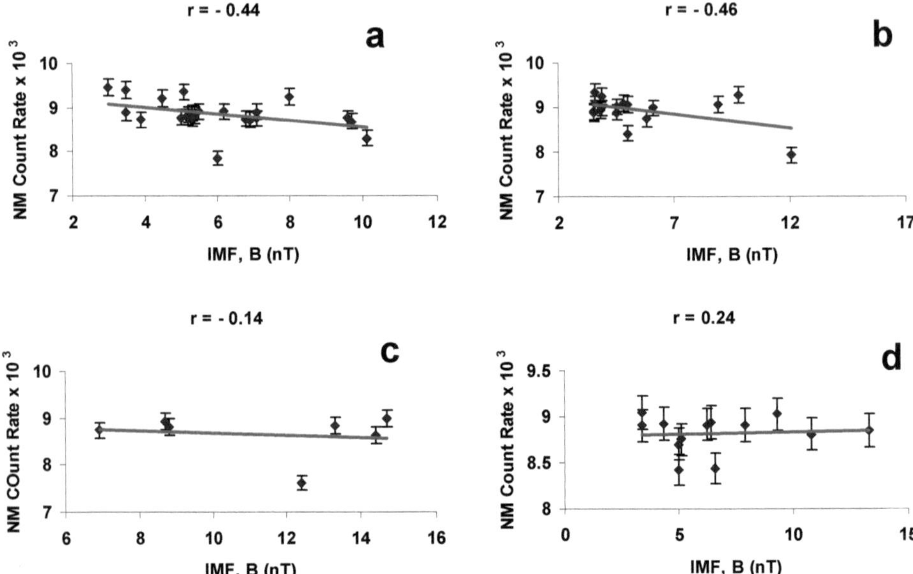
r = - 0.44
a
NM Count Rate x 10 3
IMF, B (nT)
r = - 0.46
b
NM Count Rate x 10 3
IMF, B (nT)
r = - 0.14
c
NM COunt Rate x 10 3
IMF, B (nT)
r = 0.24
d
NM Count Rate x 10 3
IMF, B (nT)

Fig 4